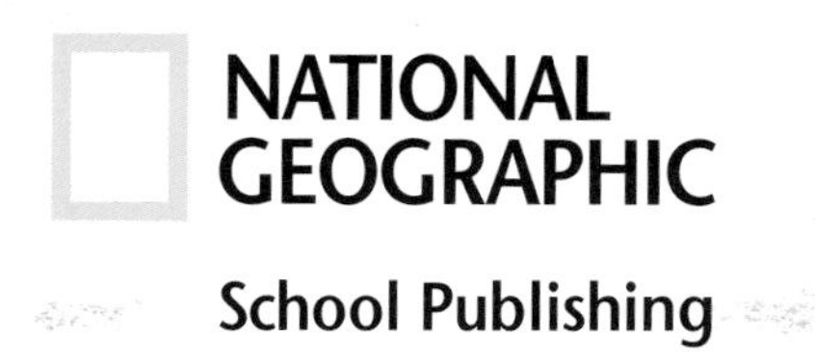

Hurricane Hunters

PIONEER EDITION

By Beth Geiger

CONTENTS

Hurricane Hunter

These fearless scientists fly into fearsome storms.

By Beth Geiger

IR FORCE
0972
815TH WRS

It was a hot September day in 2002. Captain Chad Gibson got into a plane. He strapped on his seat belt. He knew things were about to get rough.

Hurricane Lili was moving toward land. The storm was causing bad weather. And it was getting stronger. Yet Gibson was about to fly straight into the storm.

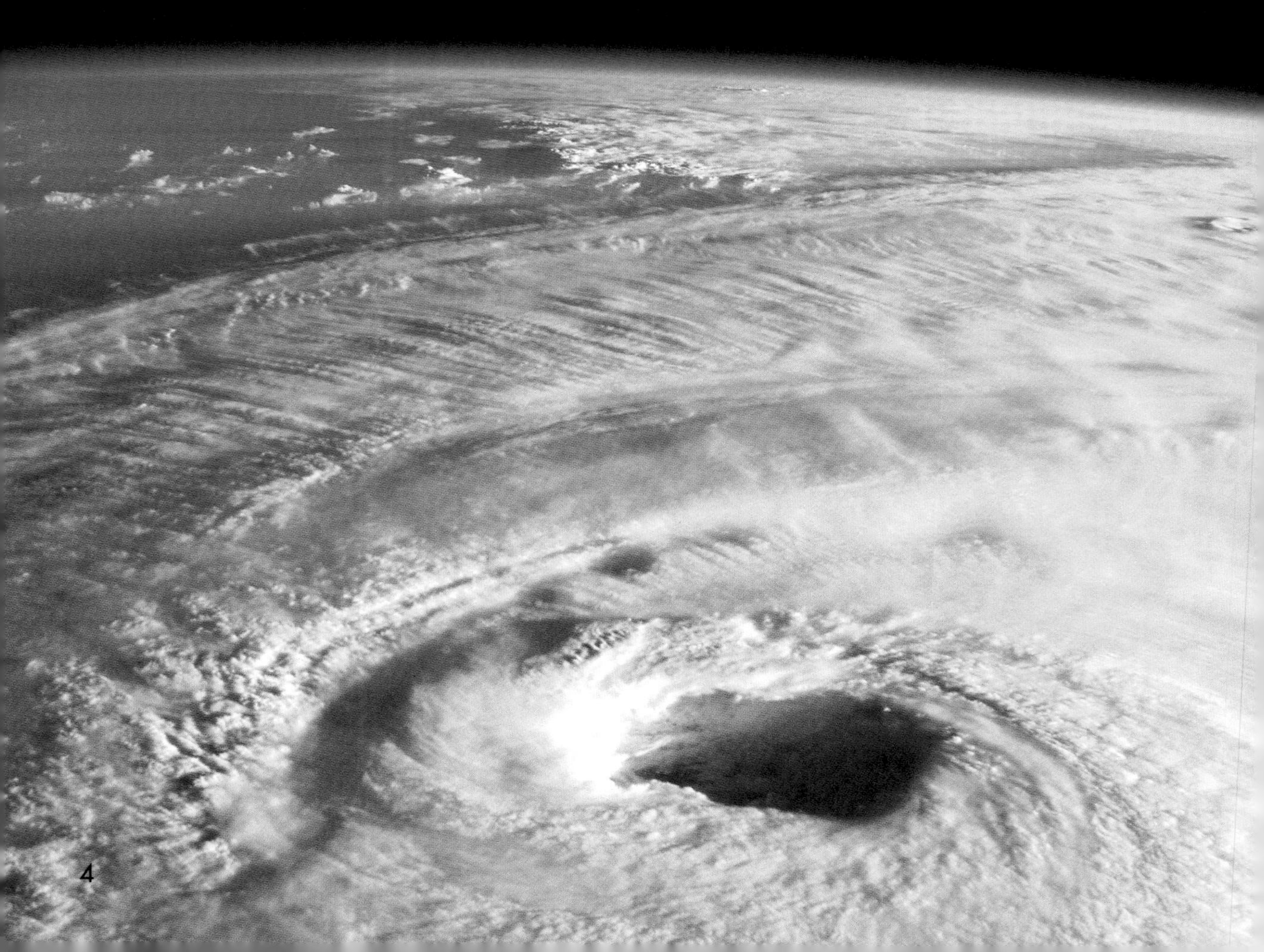

Good to Go. LEFT: Hurricane Hunters get ready to fly into a hurricane.
FAR LEFT: From space, this hurricane looks calm. Yet its winds are blowing at more than 119 kilometers (74 miles) per hour.

Hurricane Heroes

Flying into a hurricane is dangerous. The flight is bumpy. Strong winds toss the airplane through the sky.

Yet Captain Gibson knows what he is doing. He is in the U.S. Air Force Reserves. He is part of a special team. They are known as the Hurricane Hunters.

Hurricane Hunters fly into storms. Then they collect **data**, or facts. They find out the speed of the winds. They measure the temperature of the air. This data helps the scientists learn about the hurricanes.

Big Storms

Hurricanes are the most powerful storms around. They can be hundreds of miles across.

A hurricane does not start out big. It often begins as a group of small storms. These storms form over warm ocean water. They join together. Their clouds begin to twist in a circle. The winds get faster. They blow at more than 119 kilometers (74 miles) per hour. Now the storm is a hurricane.

Trouble Hits

Most hurricanes stay over the ocean. But some head straight for land. That is when they cause big trouble.

Hurricanes can blow down homes. They can kick up huge ocean waves. They can cause floods.

Hurricane Hunters follow the storms. They find out if a hurricane is headed to land. If it is, they send out warnings. They work to keep people safe.

Flying Into Lili

That is why Gibson wanted to study Hurricane Lili. The storm was headed for the United States. How strong was it? Where would it hit land? The Hurricane Hunters had to find out.

Gibson flew into the storm. Soon his plane crashed into Lili's **eye wall**. That is the ring of clouds around the center of the hurricane.

Winds rattled the plane. Lightning flashed. Rain poured down. Yet Gibson and his crew kept working. They collected data as they flew through the dangerous storm.

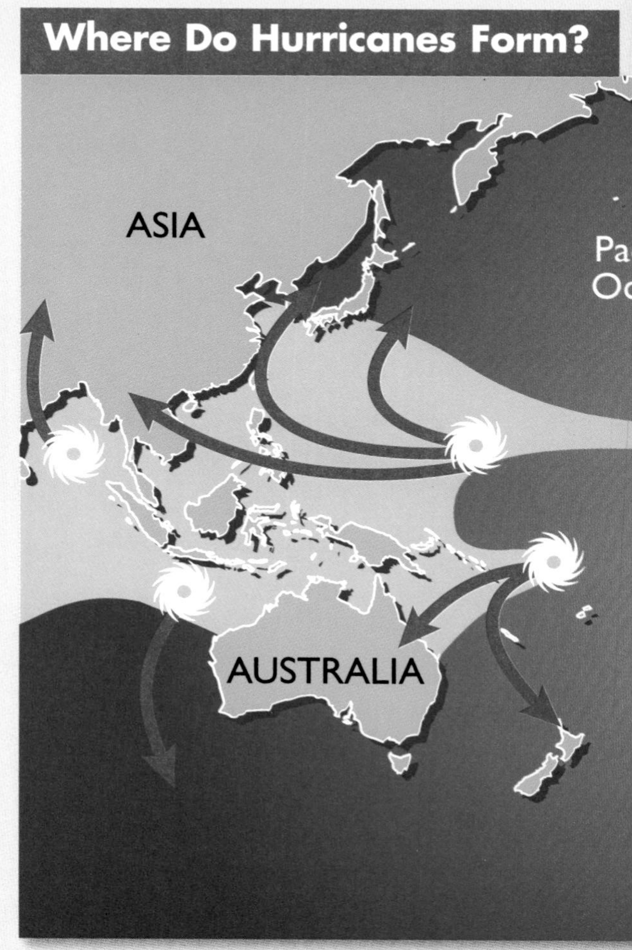

Strong Winds. Hurricane winds can knock down trees and houses.

Eye of the Storm

Suddenly, the winds stopped. Gibson could see sunshine and blue sky. He had reached the **eye** of the hurricane. That is the calm center of the storm.

It takes only about five minutes to fly through the eye. Soon the plane would dive back into the storm clouds. So Gibson and his crew worked fast. They checked their data while the flight was smooth.

After the Storm

Gibson got the facts he needed. Lili was strong. Her winds were faster than 177 kilometers (110 miles) per hour! She was headed for New Orleans, Louisiana.

So Gibson sent a warning. He told people to leave the city. The next day, Lili hit. Thanks to the warning, no one was hurt.

Wordwise

data: facts

eye: center of a hurricane

eye wall: ring of clouds surrounding the center of a hurricane

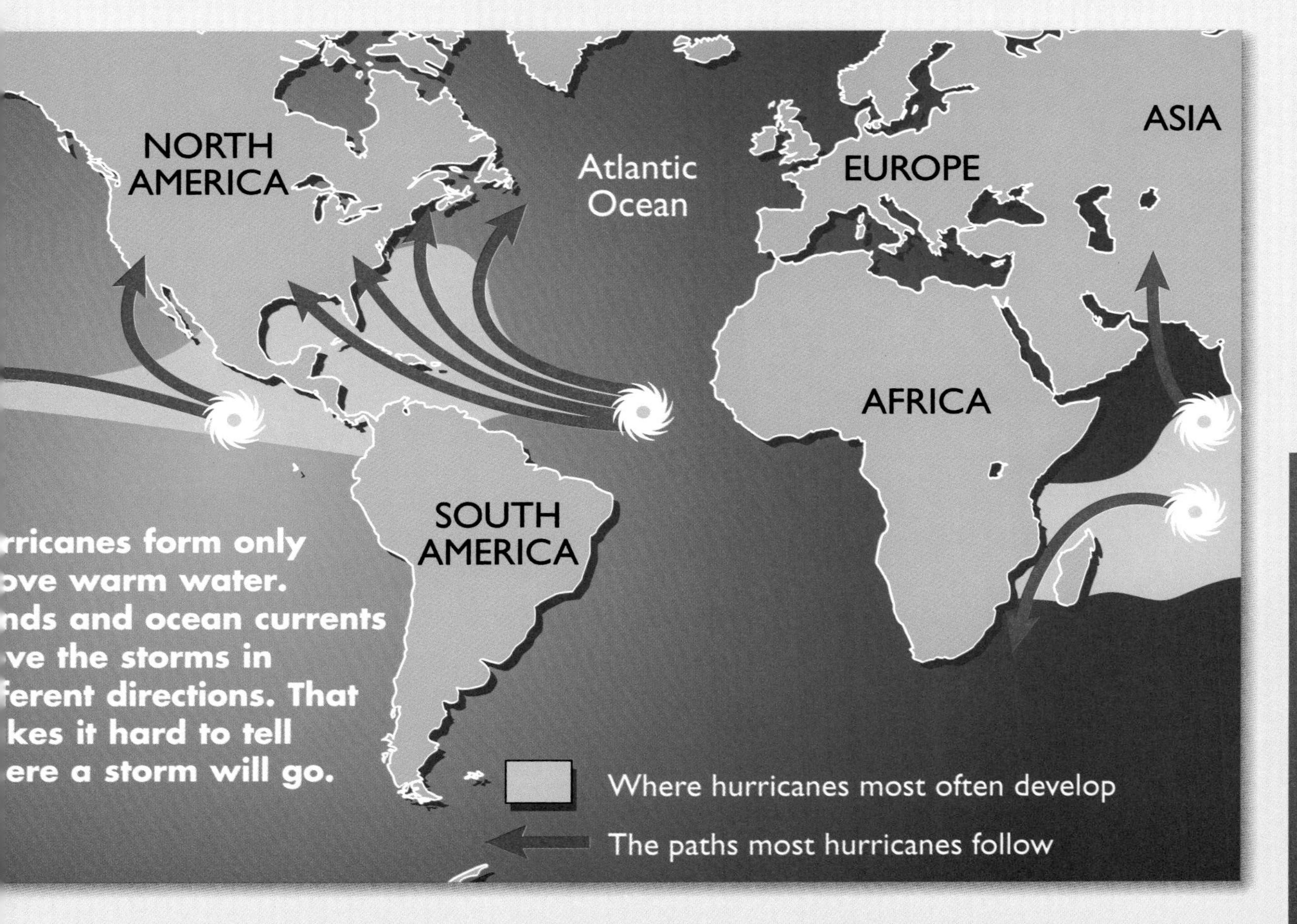

A Mighty Wind

Hurricanes are the largest storms on Earth. Yet they are hard to study. The inside of a hurricane is hidden by clouds. High winds and waves make it hard to get close to the storms. Scientists fly into hurricanes to see what is happening. This drawing shows the hidden parts of a hurricane.

Scaling Hurricanes

Weather forecasters use a scale to measure hurricanes. The scale helps them predict the amount of damage a hurricane might cause.

Category	One	Two	Three	Four	Five
Damage	Minimal	Moderate	Extensive	Extreme	Catastrophic
Winds *(in miles per hour)*	74–95	96–110	111–130	131–155	Over 155

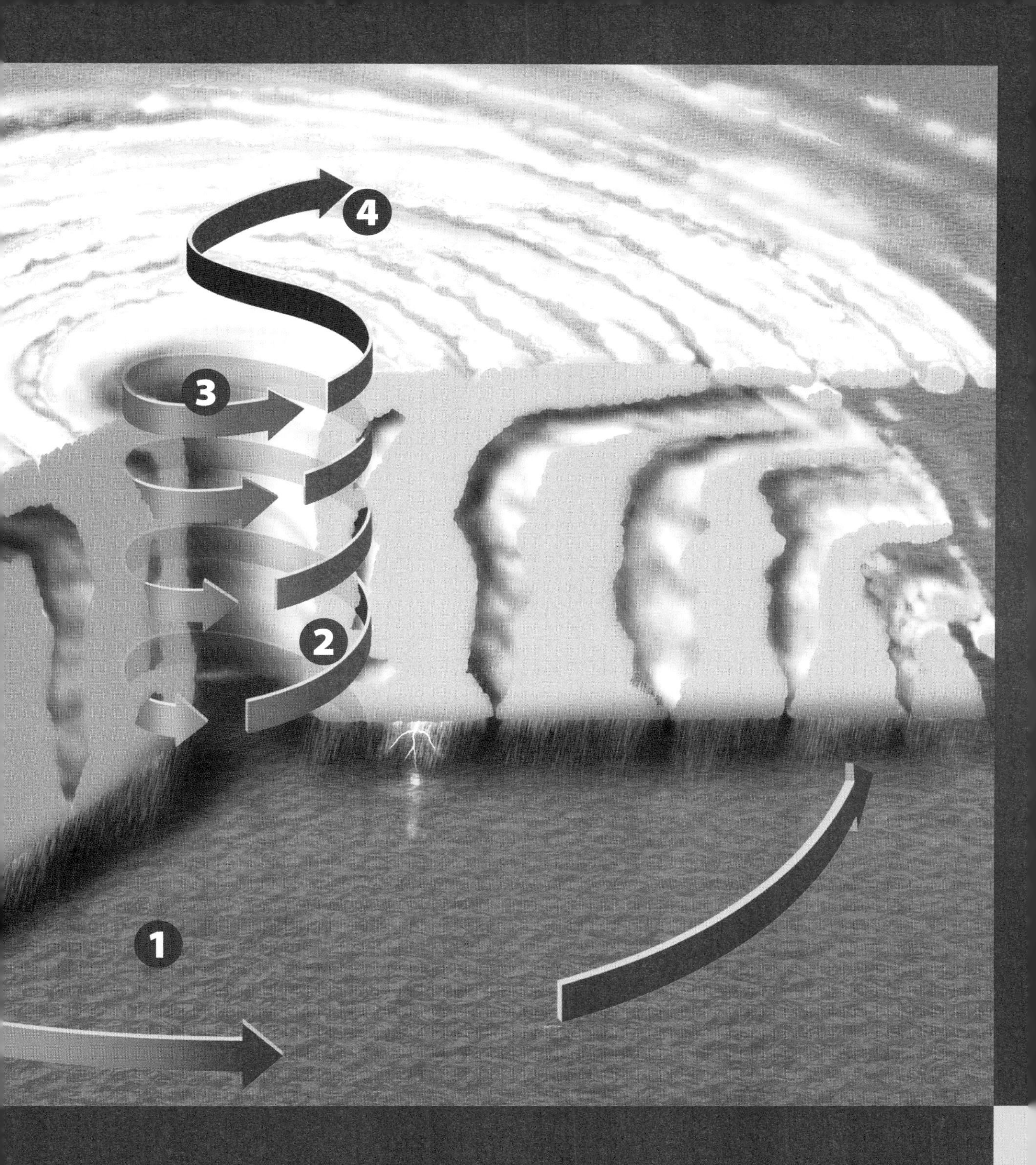

1 Winds blow toward the middle of the storm.

2 The eye wall has the highest winds. Winds can reach 322 kilometers (200 miles) per hour.

3 The eye of a hurricane can be over 24 kilometers (15 miles) wide.

4 Rising winds move away from the storm.

Katrina Hit

Each year, hurricanes hit the United States. Most cause little damage. Yet sometimes a storm crushes everything in its path. That is what Hurricane Katrina did. She hit the United States in 2005.

Storm Warning

Katrina started as a weak hurricane. But she soon picked up speed. Her winds blew faster than 274 kilometers (170 miles) per hour.

Hurricane Hunters watched the storm closely. They saw Katrina turn toward land. So they sent out warnings. They told everyone in her path to leave their homes.

Most people did. They escaped in cars and buses. They drove to places far from the storm. But some people had no way to leave. Others thought they would be safe in their homes.

Katrina Strikes

Katrina hit land with deadly force. She knocked down buildings. She flooded whole cities.

That is what happened in New Orleans, Louisiana. Katrina flooded most of the city. Many people were still there. More than 50,000 people were trapped by floods.

Rescue crews came to the city. They spent months helping people. Katrina forever changed New Orleans and the other places she hit.

Surviving the Storm

Katrina shows us why Hurricane Hunters are important. They send out warnings before a hurricane hits. That gives people time to escape.

Katrina was deadly. But she could have been worse. Without warnings, millions of people could have died.

A Rooftop Rescue. Many in New Orleans, such as this man and his dog, were trapped on rooftops after the storm.

Hurricanes

Take a spin at these questions to see what you have learned.

1. Why is a hurricane dangerous?
2. Why do Hurricane Hunters fly into storms?
3. How do warnings keep people safe?
4. Describe the eye of a hurricane.
5. Would you like to be a Hurricane Hunter? Why or why not?

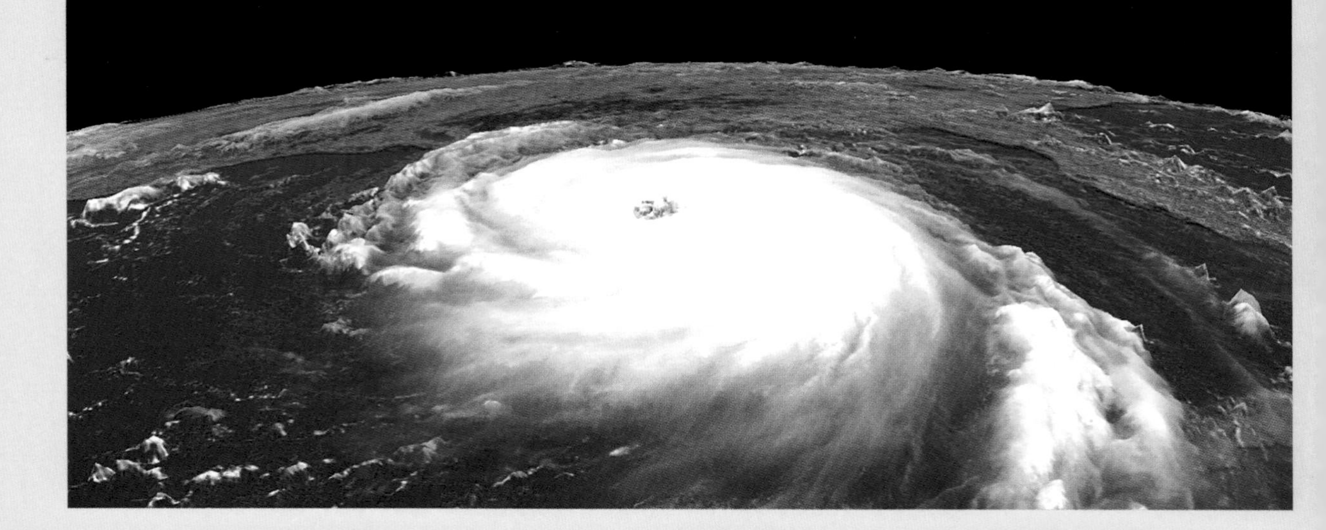

INDEX

Acknowledgments
Grateful acknowledgment is given to the authors, artists, photographers, museums, publishers, and agents for permission to reprint copyrighted material. Every effort has been made to secure the appropriate permission. If any omissions have been made or if corrections are required, please contact the Publisher.

Photographic Credits
Cover NOAA; 2-3 Ed Williams, K-Mar Industries; 4 Scott Dommin; 4-5 NASA; 6 Annie Griffiths Belt, NG Image Collection; 11 Michael Ainsworth/Dallas Morning News, Corbis; 12 NOAA.

Illustrator Credits
6-7 NG Maps; 8-9 Precision Graphics

Program Authors
Kathy Cabe Trundle, Ph.D., Associate Professor of Early Childhood Science Education, The Ohio State University, Columbus, Ohio; Randy Bell, Ph.D., Associate Professor of Science Education, University of Virginia, Charlottesville, Virginia; Malcolm B. Butler, Ph.D., Associate Professor of Science Education, University of South Florida, St. Petersburg, Florida; Judith Sweeney Lederman, Ph.D., Director of Teacher Education and Associate Professor of Science Education, Department of Mathematics and Science Education, Illinois Institute of Technology, Chicago, Illinois; David W. Moore, Ph.D., Professor of Education, College of Teacher Education and Leadership, Arizona State University, Tempe, Arizona

National Geographic School Publishing
Hampton-Brown
www.NGSP.com

Printed in the USA.
National Graphic Solutions, Appleton, WI

ISBN-13: 978-0-7362-7803-4

11 12 13 14 15 16 17 18 19
10 9 8 7 6 5 4 3

Product #4E90219

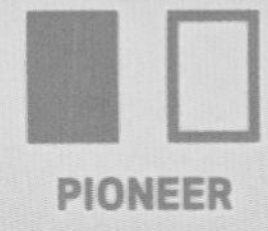

Life Science
PATHFINDER EDITION

NATIONAL GEOGRAPHIC
Explore
ON YOUR OWN

THE BEAT GOES ON

By Nancy Finton

It's Time to Explore on Your Own!

Good readers use multiple strategies as they read on their own. Use the four key reading comprehension strategies below:

PREVIEW AND PREDICT

- Look over the text.
- Form ideas about how the text is organized and what it says.
- Confirm ideas about how the text is organized and what it says.

2 MONITOR AND FIX UP

- Think about whether the text is making sense and how it relates to what you know.
- Identify comprehension problems and clear up the problems.

3 MAKE INFERENCES

- Use what you know to figure out what is not said or shown directly.

SUM UP

- Pull together the text's big ideas.

Remember that you can choose different strategies at different times to help you understand what you are reading.